WORLD DISASTERS!

FIRE

BRIAN KNAPP

STECK-VAUGHN
LIBRARY

Austin, Texas

Published in the United States in 1990 by Steck-Vaughn Co., Austin, Texas, a subsidiary of National Education Corporation.

© Earthscape Editions 1989
© Macmillan Publishers Limited 1989

First published in 1989 by Macmillan Children's Books
A division of Macmillan Publishers Ltd

Designed and produced by Earthscape Editions, Sonning Common, Oxon, England

Cover design by Julian Holland

Illustrations by
Duncan McCrae and Tim Smith

Printed and bound in the United States.

1 2 3 4 5 6 7 8 9 0 LB 94 93 92 91 90

Library of Congress Cataloging-in-Publication Data

Knapp, Brian J.
  Fire/Brian Knapp.
    p. cm. — (World disasters)
  "First published in 1989 by Macmillan Children's Books"
T.p. verso.
    Includes index.
    Summary: Discusses the causes of various types of fires, how they spread, and some of the major fire disasters.
    ISBN 0-8114-2377-8
    1. Fires—Juvenile literature. 2. Fire—Juvenile literature.
[1. Fire. 2. Fires.] I. Title. II. Series.
TH9448.K63 1990
263.37—dc20   89-11423
                                                      CIP
                                                       AC

Acknowledgments

The publishers wish to thank the personnel of the Royal Berkshire Fire and Rescue Service, England, and Steven Codrington, Australia, for their invaluable assistance in the preparation of this book.

Photographic credits

t = top b = bottom l-left r = right

All photographs are from the Earthscape Editions photographic library except for the following: title page, 6b, 8, 9, 31t, 31b, 36l, 36r, 37t, 45, LFCDA; contents page, 15t, 15b, 16tl, 16br, 17, 24b, 34br, AOIS (London); 4, 26tl, 26bl, 26br, 27 Fire Research Station, 11, 12, 30 The Museum of London; 21b, Statens Filmsentral; 13, 14, 34tl ZEFA, 40, 41 USDA.

Cover: © Ted Wood/Picture Group
Forest fire rages at Kenyon Village in
Yellowstone National Park, August 30, 1988.

Note to the reader
In this book there are some words in the text that are printed in **bold** type. This shows that the word is listed in the glossary on page 46. The glossary gives a brief explanation of words that may be new to you.

Contents

Introduction

Fires are the most common cause of disaster. There are now five billion people in the world, any one of whom could drop a match, put down a cigarette carelessly, overheat cooking oil, or clean out a gasoline can near an open flame, igniting a fire that could lead to disaster. There are also many natural processes, such as lightning and strong sunshine, that start fires. No one knows exactly how often a fire is started accidentally in the world, but it is probably a few times each second. In the United States alone, a fire is reported every 43 seconds.

What is fire?

Fire is a simple chemical reaction involving oxygen from the air, materials that will burn (called **combustible materials),** and a source of heat. Combustible materials that are commonly burned for the heat they give (such as wood, coal, or oil) are also known as **fuels.** The source of heat does not have to be an actual flame.

As combustible materials are heated they give off gases called **vapors.** All vapors have a temperature at which they will catch fire in the presence of air. This temperature

▼   *It does not take long for a fire to become a blazing inferno. This photograph was taken only three minutes after the couch caught fire.*

is called the **flash point.** In a bonfire for example, it is not only the wood itself that is alight, but also the vapors given off when the wood is heated.

Fires, good and bad

Fires are frequently started by natural causes. Lightning may set fire to trees or grassland during a storm, and the intense heat from erupting volcanos can be enough to set any nearby trees on fire. Fires are also often started by people.

Fires in nature can be helpful, burning off old dead materials from a forest floor and releasing the chemical foodstuffs (**nutrients**) that let new plants grow. A fire started by people can provide warmth, cook food, and give off energy for many useful purposes. However, both natural and human fires can get out of control. A fire turns into a **disaster** when it disrupts the normal lives of people, destroys homes, forests and other land, or kills people.

▼   *This diagram shows the way a bonfire works. Each of the features can also be found in fires that cause disaster.*

Many fires are not discovered until it is too late. The most dangerous fires are those that occur in places where people are not being careful. Many fatal fires, for instance, flare up during the night when people are asleep. Many fires also smolder for a long time, often sending out fumes that use up the oxygen in the air, so that people cannot breathe. People who die in fires are rarely burned to death; usually they are **suffocated.** At night this can happen without them even waking up.

Fires are enormously costly and insurance companies pay out billions of dollars each year for the repair of damaged buildings. The cost of fires is greater than that of any other disaster.

What makes a fire take hold?

To understand what makes a fire take hold, look at the way a bonfire is made. The first and most important ingredient in a successful bonfire is a supply of small pieces of paper, thin sticks, and twigs. These are placed in the center of the bonfire. This is the **fuse** of the fire. The small, thin, dried materials have a large surface area, allowing

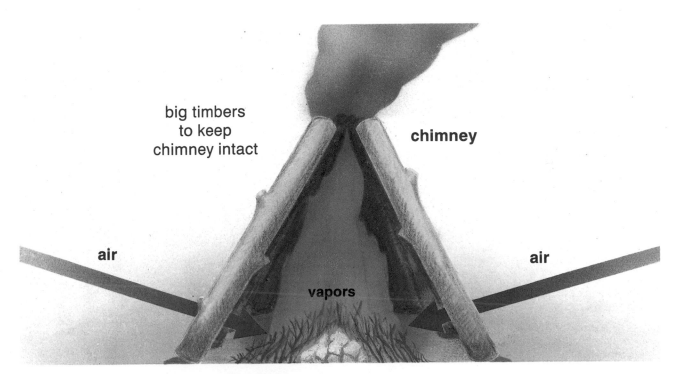

big timbers
to keep
chimney intact

chimney

air

air

vapors

fuse

▲ *This garden incinerator has wire mesh sides to allow air to get at the combustible material.*

the maximum amount of air to reach each piece. They are also packed loosely so that a free flow of air will reach the burning pieces. By this means, a continuous supply of oxygen reaches the fire.

The main part of the bonfire is built around the fuse. Larger timbers are rested against one another to make a cone that is very stable and will not collapse easily. A stable bonfire is needed to provide a natural tube, or **chimney,** that pushes hot air from the top of the fire and pulls in new air full of oxygen at the bottom. When the fuse of papers and small sticks catches fire the heat will cause air to rise. New air with more oxygen will then be sucked in at the bottom. As the fire grows the larger timbers will catch and soon the bonfire will be ablaze.

As you read through the book try to compare each disaster with this bonfire. Ask yourself: Was there some form of chimney? What provided the fuse? How did enough air get to the fire to make it so fierce? Nature and people have set up many structures just like a bonfire, although they are not always obvious until a disaster occurs.

▶ *In this factory fire the chimney was provided by an open door on the ground floor and open windows on the floor above.*

Fire disasters in the home

It is all too easy to cause a fire disaster. Here is one way it has been known to happen. A house with a large family living in it has old wiring that has not been replaced for many years. The material that keeps the wires from touching anything (**insulation**) is starting to break down. The system should be safe because all wiring is protected by electrical fuses. These are small round plugs that burn out when a line overheats. In more modern homes these fuses have been replaced by circuit breakers, which are much safer. A person only needs to flip a switch to make them reconnect.

One day a line overheats and a parent decides to bypass the fuse by inserting a penny or nickel under the fuse and screwing it in again. Now the line can be used again, and family members plug in several appliances. However, the supply cable cannot cope with the extra power being used and it heats up, melting the cable coating. Inside the walls of the house that cable is now an electrical fire; anything that touches it will smolder. In this way a fire can start in many hidden places and go unnoticed. During the night fire may start in the walls while the family is asleep and a disaster has occurred.

▼ *This sketch shows some ways a fire may occur in a home. Here a fuse has been bypassed by inserting a penny, and a socket is overloaded with appliances, causing a fire to start in the wall cable.*

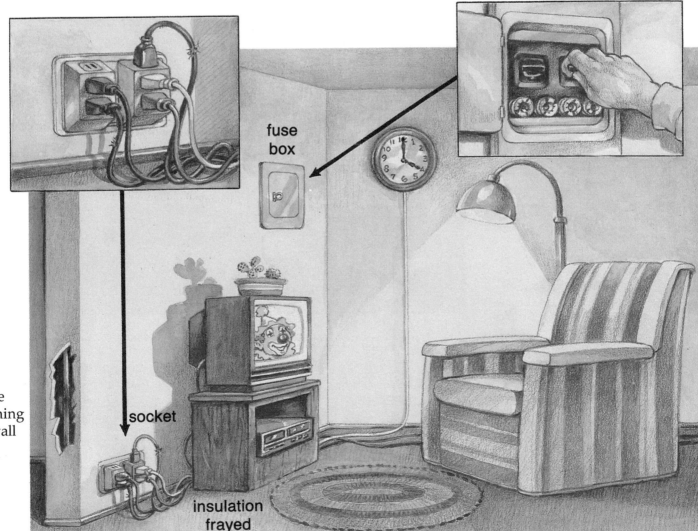

fuse box

cable burning in wall

socket

insulation frayed

▶ *A truck carrying inflammable liquids was involved in an accident. As you can see, the fire engulfed not only the truck, but also the area over which the liquid spilled.*

Fire disasters at work

Fires are found in nearly all workplaces. Some sources of fire are very obvious. In a steel factory, for example, the molten steel is at a temperature over 1,470°F. But there are many workplaces where people do not realize there is a risk of fire. In an office, for example, there are boilers that run the central heating systems. Boilers are often driven by gas or oil burners.

There is more risk of disaster in some places than in others. In a chemical plant the risk of fire disaster is enormous. Some chemicals will catch fire explosively just by coming in contact with the air. Many are extremely inflammable, and will burn fiercely if they come into contact with any source of heat at all. It is not always possible to prevent all leaks in a chemical plant and someone smoking a cigarette near such a leak could cause a major explosion.

Some places are at risk even though they may appear to be very safe. Who, for instance, would think that rock could catch fire? In coal mines there is often a natural build up of gas, and a lighted cigarette here could send a wall of flame rushing round the mine shafts in a few seconds. The gas could then set the coal on fire causing a fire that might be uncontrollable and could burn for years.

Fire disasters in the landscape

One of the worst disasters any country can suffer is when large areas of forest or grassland catch fire. Sometimes the wind fans the flames, carrying lighted **embers** faster than the firefighters can control the situation. A forest or grassland fire out of control is a terrifying experience. The fire can advance at the speed of an express train and the firefighters may get trapped. In the Mediterranean region alone, as many as 2,000 forest fires occur each year destroying over 60,000 acres of forest and scrub.

The effect of wind

Wind is the main factor that spreads fires and causes disaster. If you think of how a chimney works in a house, you will recognize why. The chimney is not built just to take away fumes; it also draws air across the fuel, bringing in fresh oxygen to keep the fire burning, just as in the bonfire already described. A building with windows and doors open can become a kind of chimney, acting in the same way, and the fire can even produce a wind of its own.

Most widespread fires are fanned by strong winds. In a forest or on grassland, the wind carries burning embers high into the air as well, floating them thousand of feet ahead of the surface fire. This makes the fire spread rapidly.

▼ *It took only a few minutes to destroy the interior of this storage room. The contents are now completely unrecognizable.*

Cities on Fire

Some of the greatest fire disasters in history have occurred in big cities. Two of the largest fires happened in London and Chicago.

In 1666 one of the world's largest fires started in a baker's shop in Pudding Lane, London, near London Bridge. Around midnight one Sunday in early September, Thomas Farriner found that the woodpile he used for stoking the bread ovens had caught fire. Unable to put out the fire and finding he could not escape from the ground floor because of the flames, the baker made his escape over the roof.

Why the disaster occurred

London in the 1600s was very different than the city of today. Many of London's houses were really slums. They were built entirely of wood, and were often roofed with thatch of rushes or straw. All the houses were huddled together so closely you could lean out of an upstairs window and shake hands with the person who was in the house across the street. There was no public water supply, no water from pipes, and no real fire department. When there was a fire, people had to put it out by themselves.

Everyone dreaded a fire because they all knew the risk of it spreading was high. There were a few primitive fire engines, but these were only useful for small fires. When a fire occurred everybody in the area left their jobs, and went straight to the scene of the fire, where they formed **bucket brigades.** These were long lines of people ready to pass bucketfuls of water from the river and local wells along to the burning buildings.

▼ *This is how the narrow London streets may have looked at the start of the Great Fire.*

Why the fire spread

The summer of 1666 had been hot and rainless. By September everything was **tinder** dry. Then a stiff wind started blowing. When the fire started in Pudding Lane close to Fish Street, the wind fanned the flames and the fire spread swiftly. One shower of sparks dropped onto straw in the yard of the nearby Star Inn. The straw flared up like a bonfire. It quickly set fire to storehouses and sheds. One by one the buildings caught fire until everything along the river front was ablaze.

Imagine the scene as the bucket brigades quickly formed lines down to the river. However, in the rush to pass the full buckets half the water spilled. The amount of water that reached the fire made no impression at all.

People near Pudding Lane had little time to think as the fire leaped from house to house. The balconies generally caught fire first then the flames crept under the roofs. Sometimes a shower of tiles or a bundle of burning straw would fall to the ground, showing how far the flames had

▲ *This painting gives an impression of what the fire looked like as it reached its greatest height. Notice the people trying to escape onto boats in the foreground. London Bridge has also started to catch fire. The large building in the center is the original St. Paul's.*

reached. People ran with screaming children in their arms, some staggered along the streets with their belongings. Others were wringing their hands hopelessly, crying out in dismay as each new house caught fire in turn. The **City Watch** were also there, trying to guard the streets leading to Pudding Lane so that thieves could not take advantage of the confusion and start looting.

At the height of the fire

By six o'clock the next morning the greater part of Fish Street was in flames and the church of St. Magnus had fallen. The houses on London Bridge were already burning, and it was obvious that the fire was totally out of control. The bucket

11

► *Firefighting equipment at the time of the Great Fire was very primitive. The most sophisticated machinery was a fire engine similar to the one shown in this engraving. The fire officer stood on the large tub and directed the jet of water at the fire, while his assistants worked the pump. The tub had to be filled by other people bringing bucketfuls of water.*

brigades had broken up and gone to save their own belongings. Thousands fled for their lives, heading from the city toward the surrounding countryside, either the marshes at Lambeth or the fields at Millbank; the Lord Mayor of London was found wandering in the streets crying helplessly. The city kept burning. That night the people on the Southwark bank of the Thames could see the Great Fire spreading eastward and north away from the river. The heavy crash of falling buildings was almost constant. Occasionally a church steeple could be seen against the night sky wreathed in flames. Then there would be a mighty crash as it fell onto the street below.

The fire spread slowly but surely toward the great cathedral of St. Paul's. The heat was so intense that buildings were now bursting into flames **spontaneously** even before fire from other burning buildings had reached them. Desperate measures were needed. Sailors from ships moored in the river brought gunpowder and tried to make **firebreaks** (areas containing nothing that will burn) by blowing up whole streets on either side of the Fleet River. Despite all efforts the fire spread across the Fleet and up Holborn Hill. St. Paul's Cathedral caught fire and was burned to the ground.

The scale of the disaster

The wind blew continuously until Tuesday night when the wind died down and the fire stopped spreading. The large flames did not subside until four days after the fire had started.

The fire was so intense that it had to be left to burn itself out. Fortunately very few people were killed by the fire but 80 percent of the city's buildings were destroyed, including a total of 13,000 houses covering 400 acres. A total of 87 churches and the cathedral were in ruins and three-quarters of the city's population were homeless.

After the disaster

Now urgent action was needed. London was the capital and very important to the country. Soon a bold plan was decided on. One of the world's most famous architects, Sir Christopher Wren, was commissioned to design a new cathedral and also asked for suggestions regarding the rebuilding of the city. The magnificent cathedral seen today is the result of reconstruction after the Great Fire. Wren also planned for wider streets. In some places his plan was followed, and there are boulevards leading to the cathedral. Nevertheless, in many places people simply rebuilt as quickly as they could using the same street layout as before. So a modern map of London still shows many of the same narrow streets that were there centuries ago.

Chicago burns down

In 1870 Chicago was a rapidly growing city built on the shore of Lake Michigan, and profiting from the cattle and grain trade of the Midwest. Although the city streets were not as closely packed as those of seventeenth century London, most houses were made of wood.

In the summer and autumn the sun scorches the city streets and sets up a breeze that blows from the lake onto the land. The

breeze gives Chicago its nickname of the "windy city" but it is also a recipe for a potential fire disaster.

When the fire struck on October 9, 1871, there was no established fire department and the fire raged through street after street. Although the fire was put out within a day it left almost 100,000 people homeless and caused millions of dollars in damage.

Just as with London, the citizens of Chicago turned disaster to advantage. They rebuilt their city center in stone and laid out the streets on a grid plan. New building techniques were also used and the world's first steel-framed skyscraper was built. To make sure Chicago's population didn't suffer in the same way ever again, one of the first modern fire departments in the country was set up.

◄ *The present-day central district of Chicago is called The Loop. Major fires often provide opportunities for architects to introduce new designs into the cityscape. The fire that happened inside The Loop helped set the scene for one of the world's first skyscraper skylines.*

Land Ablaze

Forest or grassland fires are among the worst disasters. Areas that suffer regularly from fires in the countryside include the south of Europe, southern Australia, southern California, central Chile, northern China, and South Africa.

Which types of land suffer most?

Fires are especially common and fierce in places where a warm wet season is followed by a long, hot dry season. A warm wet season allows plants to grow a lot. Then, when the weather turns dry, many of the plants will shed their leaves, covering the ground with tinder dry material that will catch fire easily.

The most serious fires occur where the dry season is accompanied by strong winds that blow for days or weeks on end. These winds fan the fire and drive it forward relentlessly. In the Mediterranean region for instance, strong winds can blow for months in the summer.

A strong wind will take burning embers high into the air then carry them hundred of yards ahead of the burning vegetation. When they land the embers set fire to new areas. In this way a fire can travel across the countryside at great speed and defeat those trying to fight it.

Australia and China are two of the countries that suffer most from **dryland** forest fires. The following examples show how fires move and how difficult it is to stop them.

Australia ablaze

In 1983 there was a massive **drought** in Australia. In this vast country much of the grassland on which so many livestock feed had completely dried up. The slightest spark, such as a glittering ray of sunlight shining through a discarded bottle, a carelessly thrown cigarette butt, any of these could set the grassland and forest on fire.

In February huge areas of **eucalyptus** forest caught fire near Melbourne. The eucalyptus tree contains a great deal of oil, which makes the trees explode into flames, as soon as they are set on fire. Because of this, a blazing eucalyptus forest is one of the most difficult fires to put out.

▶ *This is a bush fire out of control. Notice how the flames are all swept to the left; this indicates a strong wind fanning the blaze. The fire in the foreground has started some distance away from the main blaze, showing how difficult it can be to control a bush fire.*

► **These firemen are using a hose and drawing water from a nearby pond. They clearly face an almost impossible task on their own. Many hundreds of people must cooperate to fight a bush fire.**

▼ **This is the frightening spectacle that people see as a bush fire gets close to an urban area.**

▲ *This man sits in the charred wreckage of his Macedon home.*

That month there was a strong wind as well. This fanned the flames and drove them forward quickly. The hot air from the fire drew air up and then outward at a high level. It pulled burning twigs and leaves with it. As burning embers were carried away from the heart of the fire, they landed on unburned areas and started new fires.

It was an impossible task for the fire fighters. No sooner had they put out one fire than another was started by a lighted ember somewhere else.

Macedon is burned

The forest fire was soon raging on three sides of the city, coming ever closer to the suburbs. The nearby town of Macedon was to take the brunt of the fire. The wooden buildings gave no resistance to the flames that swept in from the forest and scrublands of the **bush.**

People had also brought disaster on themselves. They had not taken the advice of officials to clear every last piece of **brush** from the ground near their homes. "If fire starts" they had been told, "the brush will go up like a torch and take your houses with it." That Wednesday the terrific heat caused building after building to burst into flames spontaneously, and there was little the firefighters could do about it.

The next morning much of Macedon was a burned out shell. Brick chimneys were all that remained of what had once been well-maintained family homes. In the street were rows of cars. Now they were just burned out metal.

Fortunately, the toll of the fires in Australia in terms of dead and homeless was relatively small for such a major disaster. Seventy five people lost their lives

▼ *In Australia many buildings are built of wood, with sheet metal roofs. When a fire sweeps through a town most of these buildings are destroyed.*

and another 8,000 were made homeless. However, the area scorched was enormous. Over 815,000 acres of forest and crops were lost.

A forest fire in China

In 1987, a huge forest fire began in the northeast of China. It started when oil leaked out of a bush-cutting machine and caught fire. There were gale force winds blowing at the time. In the dry days of early summer the bushes were soon ablaze and the fire was instantly out of control.

Within days it was raging on a 110 mile front. Over 50,000 troops were rushed to the scene by the Chinese government as the fire threatened to become one of the worst the world has seen this century. Within two weeks the fire had gutted five towns, killed 200 people, and destroyed over one million acres of land.

Despite the huge numbers of people sent to put out the fire, they could not stop its advance toward the town of Mangui and its population of 20,000. Also at risk were another 3.7 million acres of forest surrounding the town. A force of 800 special firefighters was sent to protect this one town alone.

The Chinese went to extraordinary lengths to put out the fire. They cut a 75-mile-long firebreak to try to stem it. They also tried spraying crystals into the air from high flying aircraft to make the clouds produce rain.

The disaster from this fire was immense. Not only did it cost the equivalent of hundreds of millions of dollars to put out, but it burned away a large part of China's scarce timber reserves. It also destroyed much of the country's last area of natural forest. It is not certain whether that forest will ever fully recover.

◄ *Dry areas such as Australia, China, and California suffer from frequent fires. In many areas the fire is fanned by natural winds that occur every summer, just when the brushwood is at its driest. These winds make fire fighting even more difficult.*

Energy Crisis

Disaster from fire does not always produce dramatic scenes of land or buildings ablaze. Disaster can also come slowly and unseen.

Burning fuels may not seem a disaster at first. How can a fire controlled in a hearth be a disaster? Yet there may be nearly a billion hearths in the world. No one fire is important in itself, but together the billion domestic fires, the millions of industrial fires, and especially the giant coal-burning power stations destroy more land and cause more worldwide damage than any accidental fire.

Fire for heat

It is a great modern convenience having hot running water, electricity, and central heating at our disposal. People need to cook to eat, and many climates are too cold in winter or at night for people to survive there without some form of heating. None of these things is possible without the heat energy from fire.

► *This African woman has to collect firewood every day because she is too poor to buy oil or coal. In many parts of Africa the daily search for wood by millions of people has meant that trees are becoming very scarce.*

In the **developed countries** people burn coal, oil, and gas, the **fossil fuels.** They are burned in power stations to produce electricity, or in our furnaces, boilers, and stoves at home. This is gradually using up the supply of these fuels. When they are gone the world may face an enormous disaster. Some will be gone in just a few decades. There is already a disaster much nearer at hand, however. It does not affect people in the industrial world, but those who live in the **developing countries,** and who make up the majority of the people on this planet.

Burning wood to survive

Burning wood for fuel and to make way for more farmland is causing a disaster today. It is forcing people to use up the world's forests faster than they are being replanted. This is because all over the developing world the main source of fuel is wood. Firewood is used by the poor in developing countries for cooking, to heat water for

▲ *This man is putting the finishing pats of mud to the outside of a brick kiln. Sticking out from the bottom of the kiln you can see whole tree trunks. Enormous quantities of wood are used in making bricks for houses.*

washing, and to keep warm. Wood is burned to bake bricks in a kiln and to smelt iron. The demand for firewood is huge. There may be four billion people who depend on wood as their source of fuel.

In developing countries people are burning wood just to clear the land. In some tropical countries whole forests are being cleared to make room for more farmland to feed the population. Some of the cleared land will be good for farming, but most of it is not very fertile. However, the burning of the forests continues.

Although these fires are all well controlled, burning forests on this scale is an enormous disaster. An area the size of Massachusetts has its forest burned away each year.

▶ *The beautiful Himalayan mountains used to have forests covering their lower slopes. Now most of the slopes are bare because the trees have been cut down for firewood. Without the tree roots to help water sink into the soil, storm rains flow quickly to rivers and cause flooding disasters.*

Burning the forests is a waste of timber that could be used in many ways. It could be used to make furniture, buildings, and paper; it could be used to cook innumerable meals and keep large numbers of people warm. Instead it just goes up in smoke.

Making deserts

People are cutting down trees for their fires but there is little replanting. You can see the results everywhere. The majestic Himalayan mountains were once covered with woodlands. Only fragments now remain. Goats graze on land that once bore trees. The search for firewood goes on throughout Africa. As each new tree is cut down and burned, there is less protection for the land and the soil is more easily washed or blown away. The process is called **desertification.** It makes the soil less fertile and makes it difficult to get good crops.

◄　　*Although we cannot see it happening, this power station releases thousands of tons of sulfur gas and carbon dioxide into the air every year. This helps to cause acid rain pollution and helps to change the world's climate.*

Changing the climate

Fire is a chemical process. When fuels are burned they give out heat, but they also produce smoke (carbon particles in a gas) and a large variety of gases. One of these is **carbon dioxide**. Carbon dioxide has been pouring into the air for many centuries, but the amounts are increasing dramatically year by year.

Carbon dioxide in the air absorbs some heat from the Earth that would normally go out to space. It is slowly making the Earth a warmer place to live. This is not really good news, since it also means that climates are changing all over the world. It seems likely that even within our own lifetimes large parts of the world will get drier as the climate changes. These are the places that currently produce most of the world's food. Burning fuels could thus lead to a food shortage!

As the world grows warmer, the ice caps in the polar regions will begin to melt. If all the world's ice melted, the level of the water in the oceans would rise by up to 260 feet. This would drown most of the world's great cities, because most are built on low-lying land near the coast.

Acid rain

Two of the most poisonous gases given off by burning fuels are sulfur dioxide and nitric oxide. As they mix with water droplets in the clouds they form weak acids. When the water falls from the clouds the droplets fall as **acid rain**.

▼　　*The pollution from fires goes into the air and makes the water droplets in the clouds acidic. When the droplets fall as rain they can cause the leaves of trees to turn brown. Eventually many trees die.*

Acid rain comes directly from burning fuels. When acid rain lands on forests, trees die. When it falls in some lakes, all the creatures in the water die. This does not come about quickly. The destruction takes place over many years. The effects of acid rain are now thought to be one of the world's great disasters.

How Nature Protects Itself

It is very dangerous for people to be near a fire that is out of control. Fires appear to consume entire areas, destroying plants and animals, and laying everything to waste, but are appearances deceptive? What is the real impact of fire on the natural world? Is fire as much of a disaster for nature as it is for people? How does nature cope with this common event? What lessons can we learn from nature?

▼ *This diagram shows the many ways trees, grasses, and animals protect themselves from fire.*

Plains and grasslands

The way nature uses fire to advantage is most clearly understood by looking at a place where fires are common. Plains covered by a mixture of trees and grasslands make up a great part of Central Africa. These plains are called **savannas.** The central regions of America and Asia also have enormous grasslands. In the United States the grasslands are called **prairies** and in the Soviet Union they are called **steppes**.

For several months of the year these regions receive no rain and the plants become tinder dry. When lightning strikes, the vegetation catches fire readily. But little long-term damage is done because the natural vegetation has the means of protecting itself.

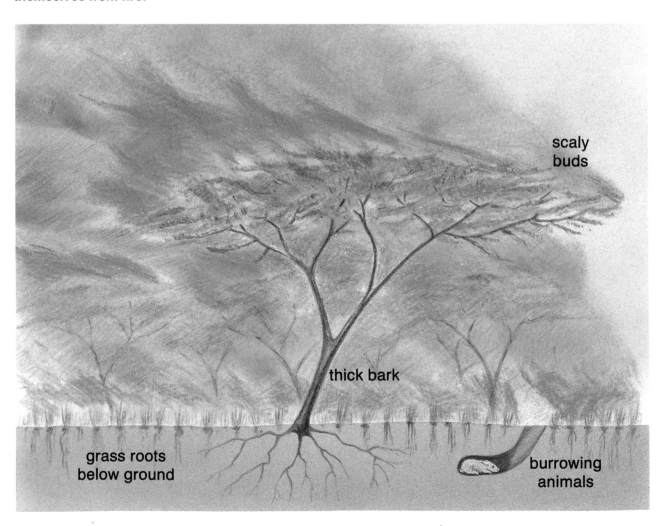

scaly buds

thick bark

grass roots below ground

burrowing animals

How grasses cope

Nature employs many kinds of protection against fire. For example, most grasses have their shoots as well as their roots buried below the ground surface. A fire that sweeps swiftly over the landscape will not scorch the soil. Other grasses have pointed seeds that dig themselves into the soil as they fall from the plants. In this way they are less likely to be scorched by a fire.

Dry grassland plants are not only protected against fire, they use fire to help them to grow. Most plants are very woody, and their dead remains do not rot easily in most grasslands because there is so little rainfall. As a result the nutrients in the dead leaves are locked out of reach of the growing plants. Fire turns the dead leaves into ash. When it rains next the ash is washed into the soil. There it releases its nutrients and acts as a natural fertilizer.

▲ *When a fire burns quickly it only singes the plants. Here you can see parts of the plants have been left untouched and still remain light brown. The soil, however, is strewn with the ash that will return nutrients to the soil when the next rains fall.*

◄ *This photograph shows fire sweeping across grassland. The fire is burning over a wide front, but it is moving quickly and large flames can be seen in only a few places.*

How trees survive

In areas where fires are common, trees have ways of protecting themselves, and like the grasses, they use fire to their advantage.

One of the most common ways of protection against fires is the development of a thick cork-type bark, around trunks and branches, that even appears on new shoots.

Tree buds are shielded against a fire by thick scales. The buds will remain unburned even if all the leaves catch fire. If a severe fire causes the buds to be burned and destroyed, most dryland trees can quickly produce new buds along their stems and on their roots. It is difficult to kill a dryland tree.

◄ *This is the thick protective bark of the cork oak tree.*

▶ *As fire burns through this savanna, flames lick at the branches of the trees. Although the branches are charred on the surface, and many of the leaves are burned away, the tree remains healthy.*

How fire helps the redwoods

The redwoods are the world's biggest living things. Their enormous trunks are about 13 feet across and they rise to 150 feet. Most redwoods grow slowly. The giant redwoods that are now such a famous part of the Californian National Parks were seeds over 2,000 years ago.

Slow growth is not an advantage to a seedling. Many other plants will grow faster and they may shade out the redwood seedlings. Without enough light the little redwoods would die. So, surprisingly, fire is essential to their survival.

▲ *Small rodents can easily survive in a grassland where fires are common because they can dig to safety.*

► *The giant redwood shown here has been scorched many times. Notice the absence of competing plants.*

A fire sweeping through the forests of California will often destroy most of the ground vegetation. A very severe fire will burn down many large trees. However, the bark of the redwood is over 6 inches thick and it will be little more than scorched. The fire will produce much ash and this will benefit any plant that comes up quickly after a fire. The redwood cones burst open a few hours after the fire is over and the seeds fall into the warm ash, ready to germinate quickly. With the ground vegetation burned away the redwood seedlings get a head start. Each time another fire starts, the competition is burned and the redwood remains.

Animal survival

Just as many trees get burned in a fire so, too, many animals perish. However, in areas that have many fires, species would be wiped out if they had no means of survival. Many larger animals can swiftly run away from a fire. Smaller animals simply dig in. The prairie dog, for example, a familiar **rodent** on the American grasslands, lives in burrows that serve as dugout shelters when a fire sweeps across the grassland.

Why People Die

People are more frightened of fire than any other force of nature. The fear of being burned is taught to children as one of their first lessons. Schools, for example, are required to carry out fire drills so many times a year. It is all the more surprising that fire is among the most common forms of disaster in the world.

How many die?

Nobody knows exactly how many people in the world die from fires because most individual fire disasters do not kill large numbers of people. It is likely that fires now kill more people than any other form of natural disaster. For example, fire kills about one person in every 25,000 in developed countries such as the United States and Australia.

There are many ways in which fires can kill. In Britain, for example, where fire kills about one person in every 50,000, about 700 die in house fires and of these 300 die from the poisonous fumes given off by foam-filled furniture. This happens despite the large number of regulations designed to reduce the risk of accidental fires. It occurs despite a well-trained and efficient fire and rescue service that remains on constant

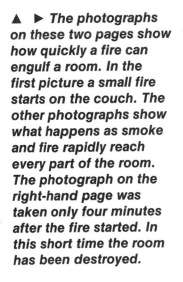

▲ ▶ *The photographs on these two pages show how quickly a fire can engulf a room. In the first picture a small fire starts on the couch. The other photographs show what happens as smoke and fire rapidly reach every part of the room. The photograph on the right-hand page was taken only four minutes after the fire started. In this short time the room has been destroyed.*

duty. It also occurs despite the large amount of research that is undertaken to find safer ways of building and less fire-prone materials.

In Britain the risk from natural fires is small because there is no long dry season, yet 1,000 people die each year. However, in many developing countries the risk from natural fires is much greater because fire regulations are at a lower standard.

Why is fire disaster so common?

To make a fire you need fuel, a supply of air, and heat. Each of these elements are readily found around the home. The trouble is that many people often do not recognize the potential danger.

For example, a discarded glass bottle left on a pile of roadside trash may act like a magnifying glass and focus the sun's rays onto combustible material that might catch fire. The person who threw out the bottle did not expect it to cause a fire. The person who discarded the trash did not think of it as combustible material. Nevertheless, a fire has started.

Small fires can lead to large ones

Imagine what might happen if a cigarette is carelessly thrown away in a boiler room on the ground floor of a building. If the cigarette lands on some old papers or oil rags it may smolder and finally set the trash on fire. A small fire started in this way could so easily become a big fire.

Because the door to the boiler room is closed and unattended, the small fire is not noticed. Inside the combustible materials are smoldering, using up all the air and creating a situation called **incomplete combustion.** Now the room is full of hot gases under pressure "waiting for air."

How people make the fire worse

The first sign of a fire might be a small wisp of smoke curling under the storeroom door. The door should be kept closed and the fire department should be called immediately. However, if the people in the room have not been instructed on how to deal with fire, they will probably open the door to see what is the matter.

The moment the door is opened a **flashover** may occur. The hot gases in the room instantly get a new supply of oxygen that causes the room to burst into flames. The whole room is immediately set ablaze and within moments the fire is able to spread out into the corridor. Sometimes the flames spread so quickly that the corridor also fills with flames and smoke in seconds.

There is no way of stopping this fire, but its effects could be slowed down. In this case, however, people have turned the building into a fire trap, because all the doors are open.

Smoke starts to travel upstairs. The front door is left wide open to make it easy for deliveries to be made. People have propped open the fire safety doors near the stairs to make it easier to enter and exit. The upstairs windows are all open.

Each of these actions has turned the building into a "chimney." People rush upstairs because the fire is filling the bottom of the building. The flames follow, sucked up by the chimney effect.

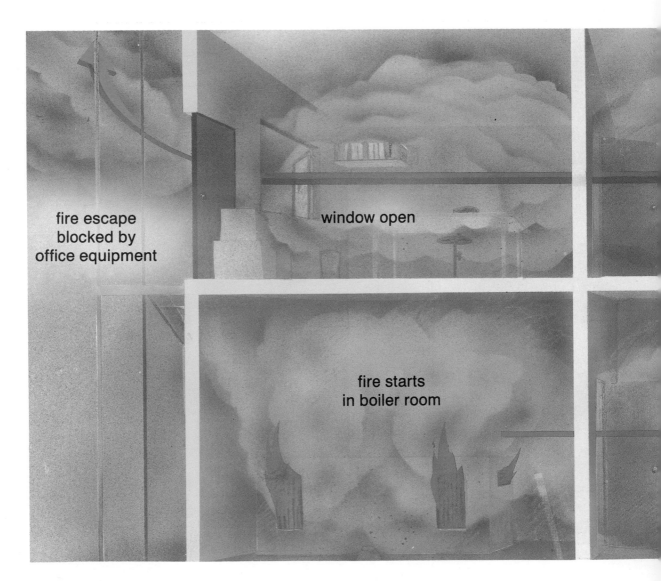

fire escape
blocked by
office equipment

window open

fire starts
in boiler room

Poor building design

One of the most famous examples of an office building fire occurred in Brazil's Sao Paulo in 1965. A fire started in one of the lower floors of a skyscraper that was not fireproof. There were no emergency staircases for the people to get out. Within two minutes the bottom ten floors of the building were ablaze.

People rushed upward to the roof, but the firefighters couldn't get to them in time. Within 20 minutes 227 people had died because not enough care had been taken in designing the building.

▼ *This diagram shows some of the many reasons a fire can take hold in a building.*

Dangerous rubbish

Disaster can strike even while people are enjoying themselves in the open air. This happened in England's Bradford City Stadium in 1985 during a soccer game.

The wooden stands where many of the spectators had crowded together were old, and beneath them a huge pile of discarded papers and cartons had accumulated. It is thought that a cigarette accidentally dropped through the stands and onto the pile of trash.

The fire caught hold very fast and with little warning. Within a few minutes the stand was completely ablaze and 56 people were killed.

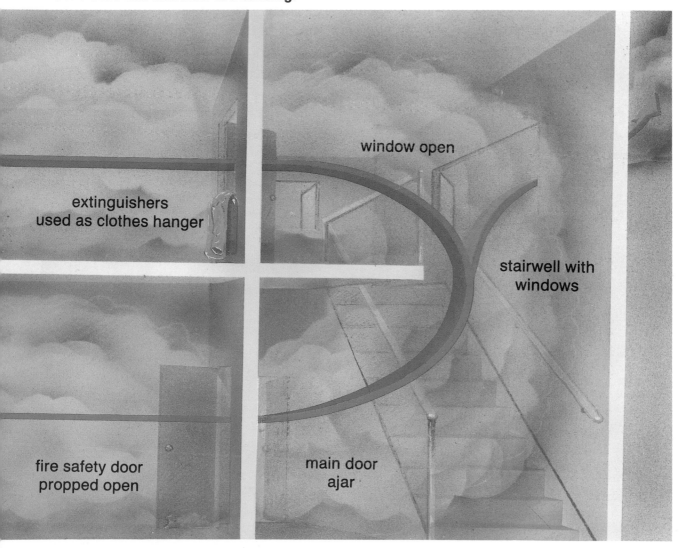

extinguishers used as clothes hanger

window open

stairwell with windows

fire safety door propped open

main door ajar

Great Disasters

Fire disasters are very common, and each one may affect only a small number of people. A single fire disaster cannot usually match the tragedy brought on by natural hazards such as drought, flood, storm, or earthquake. In some of these incidents millions of people have died. This is because natural disasters can come about without warning affecting large areas at a time.

People have rarely suffered from large scale fire in the past. However, all this was changed by the invention of fire bombs (known as **incendiary bombs**) and their use during World War II (1939-45).

The incendiary bomb

When an incendiary bomb is dropped from an aircraft, it does not destroy buildings with a large explosion. Instead it explodes in such a way that burning fragments are scattered over a wide area. Just one bomb can set a whole block of buildings on fire.

The device was first used on a large scale by the German air force in 1940 when they attacked Poland and England. Bombers were sent out from Germany to drop bombs on the great Polish and English cities. Many of the bombs were incendiary. The intention

▼ *This is a view of wartime London looking east. The Tower Bridge has been left unharmed, but the London docks are badly damaged. The pall of smoke shows the location of the dockyards.*

was to create so much damage that the fire-fighters of the cities could not cope and industry and transportation systems would be put out of action. It was thought that this would bring each country to the point of surrender very quickly.

Enormous numbers of aircrafts dropped millions of incendiary bombs, setting London ablaze night after night for week after week. Yet somehow, day after day, the fires were put out, and the shaken people of London tried to pick up the pieces of their lives. London survived, and German hopes were disappointed.

Cities destroyed

Incendiary bombs did destroy whole city centers, both in Great Britain and Germany. Many areas were affected so badly they had to be completely rebuilt after the war. That is one reason why many European cities have so many new buildings in their city centers.

▲　*Firefighters tackling the dockyards' blaze from the river.*

▼　*A building collapses after extensive fire damage.*

Some cities were attacked savagely and were very badly damaged, such as the English ports of Southhampton and Plymouth. One night the attack was focused on Coventry, the English engineering city. The city center was set ablaze and many people died.

A few years later, when the tide of the war had changed and the allied forces were beating Germany, the Allies destroyed many German cities by fire.

Perhaps the most severe attacks were made on the cities of Dresden and Cologne in 1945. Dresden was a major engineering

◄ ▼ *These three photographs show the reconstruction after the fire bombing of Cologne. The left-hand picture shows the city still smoldering in 1945. Only the cathedral spires can be identified among the rubble. Below is a picture taken in 1949 of the waterfront near the cathedral. In this scene the bridge is just being completed. On the opposite page is the same view of the waterfront today. There are several ''new'' buildings. These are reconstructions of buildings that existed before the war, and which were rebuilt using old photographs and drawings.*

city, while Cologne was the hub of much of Germany's transportation system. Large sections of both cities were set on fire with incendiary bombs.

When the bombing was over Dresden and Cologne were simply burned out shells. Dresden was almost wiped off the map. In Cologne, bridges and over 80 percent of the buildings were destroyed. So much damage was done that people couldn't even find the original road pattern because it was buried under the rubble of collapsed buildings. An estimated 70,000 people were killed in one night during a raid on Cologne.

Fire storm: the worst disaster

A city can burn so fiercely that hurricane force winds are created. This happened when cities such as Warsaw in Poland were hit by millions of incendiary bombs during World War II.

The fire was so intense that it sucked in vast amounts of air from every direction. Soon hurricane force winds blew along the city streets. A **fire storm** had been created.

At the heart of the fire no one could survive. It was not possible to breathe because the fire consumes all the oxygen, nor was it possible to take shelter from the heat. Just outside the zone of devastation the winds were so strong they blew people along the streets and toward the fires.

Napalm

Another bomb has been developed that will spread fire widely when it explodes. This is called a napalm bomb and it was used extensively in the Vietnam War. Napalm is a mixture of gasoline and other fuels, together with a chemical that makes the mixture into a jelly. Its use caused great destruction and many casualties. Many people protested about its use in any future warfare.

Emergency

A fire emergency calls for swift and expert action. The first job is to save lives. After time, although every situation is different, the principle of firefighting is always based on removing one part of the fire triangle: either fuel, air, or heat.

Emergency in the country

The major problem with a forest or grassland fire is to find a large enough supply of water to quench the flames. In desperate circumstances, firefighters sometimes bomb the fire by releasing huge bags full of water slung below planes and helicopters. This can have dramatic results, but it is very expensive.

Forest and grassland fires usually burn on such a wide front that removing the

▲　Water is dropped from a special bag suspended below the aircraft. The amount of water that can be carried by this system is small and it is used only in special circumstances.

▶　A firefighter uses a chain saw to cut down a burning tree, and thereby prevent the spread of the fire to the unaffected part of the forest.

source of heat by using water is impractical. Nevertheless, in an emergency large teams of beaters can be used to try to put out ground fires. This is especially important in areas where a fire has swept through but only burned some of the dried leaves and wood. Smoldering embers can catch fire again very easily and bring back a full scale emergency.

Emergency firebreaks

One way of stopping the advance of a forest or grassland fire is to create a firebreak—a strip of land that contains no combustible material.

When a fire looks like it is getting out of control, one team of firefighters must work a long way from the fire front, perhaps several tens of miles away. Here they use chain saws to fell trees and bulldozers to clear the ground. In some desperate circumstances firefighters even use explosives to blast a firebreak open. The wider the firebreak can be made, the better

the chances of success, but fires burn so quickly that the firefighters have to work at a frantic pace.

Once the firebreak has been formed, the land on the side nearest the fire is often set on fire. The fire can only burn one way because there is no fuel in the firebreak. It therefore burns slowly toward the major fire. As the two burn together they widen the region without fuel and increase the chances of halting the main fire.

Knowing about the weather

The person controlling the firefighting will have to make sure there are firebreaks in the direction the fire is advancing. One firebreak may not be enough. In an emergency there rarely will be time to make long firebreaks, so the best use must be made of what *can* be prepared. To do this the controller relies heavily on the advice of the weather forecasters to find out the likely direction and speed of the wind. Both these pieces of information will be needed not only to choose the direction and width of the firebreak, but also to estimate how far away the break must be started from the present fire front.

▼ *This diagram shows the way firefighters try to contain the fire, by setting up firebreaks in the direction of fire spread.*

wind

forest set ablaze here to widen the firebreak

firefighters beat flames and use hoses

firebreaks

Emergency in the city

The job of the firefighter in the country is to try containing the fire into as small an area as possible. This is also the job of the city firefighter.

It is not always easy however, because many buildings are placed close together. A fire may start in a row of houses and spread from one to the other. The heat from a large fire may make buildings and the other side of the street so hot they also burst into flame.

Emergency work in the city needs close coordination. As soon as the fire trucks arrive on the scene, the firefighters split up into teams, each with a special task. One team has to search for any trapped people. They may have to enter the flaming building using special protective clothing and breathing apparatus, or they may use extension ladders. Whichever they use, it is a very dangerous task. Sometimes even firefighters get hurt when fighting fires.

At the same time as people are being rescued, another team goes around the outside of the fire to determine the size of the problem. Knowing the scale of the fire is a vital first step in getting it under control. Yet another team must get hoses connected to **hydrants** in the street. After this, special firefighting equipment and extra apparatus

▼   *A small child who has been taken from a burning building is handed to an ambulance man. The ambulance team is another vital part of the emergency team.*

▼   *This is a fireman's view over the rooftop of a burning building. The men on the ladder have to cope with the heat of the fire, the hazard of smoke, and the danger of explosion during the first vital minutes of an emergency.*

can be called for if they are needed.

A large part of the skill in firefighting is making sure that everyone is working together. This is why a trained team is needed. Water is the chief aid in firefighting. It is used in most fires unless there are special chemicals involved where water can make a fire worse. Firefighters need to direct water at the base of the fire, and at the same time they have to use the water carefully and sparingly. If they are not careful, not only is water wasted, but the water itself can cause much damage.

▲ *Firemen with breathing apparatus tackling a severe blaze.*

◄ *This picture, taken after the fire had been brought under control, shows the complexity of a fire emergency. Notice the network of hoses in the street, the use of a turntable ladder, and the back-up trucks in the background.*

Care in the Countryside

There are many different causes of fire. In the countryside a flash of lightning can set fire to grassland or trees. A bottle, carelessly thrown on the ground, can act like a magnifying lens, and focus the rays of the sun on dry grass. Within seconds the grass can catch fire, beginning a chain of events that will lead to disaster. A truck carrying inflammable liquids can crash and catch fire on a street or highway. If the truck's load spills onto dry grass or trees a fire will spread instantly. So how do we prepare for such a wide variety of problems?

Risk of natural disaster

The first way to be prepared is to know the risk of disaster. As we saw on page 14 there are some places where natural fires are common, and in those places the risk of disaster is very high. The main areas at risk are the world's seasonally dry lands. However, many of the world's forests are also at risk in a hot summer, even in areas that normally have rain in every month. It is not unusual for some national parks and forests to be closed to the public every year due to the damage of fire.

Increasing the fire risk

The risk from natural fires is greatest where a great deal of dead brush has built up on the forest floor. This may be the result of logging, cutting away the lower branches to make trees grow straight and without "knots."

Some of the areas at greatest risk are those on the outskirts of cities, such as Los Angeles or Marseilles in France. These are countryside areas that have been invaded by people seeking a quiet life. The houses are set far apart and are surrounded with large gardens. Many of the gardens are either too big for the families to manage or they are kept with a "wilderness" look because this attracts wildlife.

"Wilderness" gardens might prove to be dangerous in a fire risk area. Dead brush left on the ground will catch fire easily. A fire will spread very rapidly and endanger lives and destroy property. The people of Macedon near Melbourne (page 16) lost their homes in just such a suburban fire.

remove small trees and brush

prune branches

pile and burn

build a fire trail

The National Park Service

The National Park Service manages some of the most valuable and frequently visited forests in the world; forests that often contain unique species of trees. These trees need to be protected from fires caused by careless visitors.

Each summer the Park Service sets fire to its own forests! The foresters have learned that fires are necessary to the forest, so the Park Service regularly burns its redwood forests to keep the redwoods healthy.

There are other reasons for setting fires. The Park Service once prevented all fires, but they found this caused even greater disasters. The reason is that if you prevent all fires in a dry climate, dead vegetation simply piles up around the base of the living trees. It is like building a bonfire.

During the summer there will be many thunderstorms, and these will include flashes of lightning. Lightning striking a forest quite often starts a natural fire. If there is little dead matter on the ground, the fire will be small and will do no damage. If it is many years since the last fire, the dead matter will burn fiercely and this may destroy the trees. By trying to prevent all forest fires the Park Service was actually making disaster more likely. As a result the forest deadwood is now burned regularly.

▼ *This forest is scorched, but it is controlled burning by the Park Service.*

◄ *The Park Service encourages people to enjoy campfires, but also to take precautions to stop fires from spreading out of control.*

On the moorlands of Britain there is also regularly controlled burning. Again, this helps to prevent any natural fires (or those caused by careless use of matches and cigarettes) from getting out of control. It also helps to release nutrients for further plant growth.

Protecting a commercial forest

In many forests it is not possible to allow any sort of burning. This is especially true in most forests of pine and spruce trees, because these trees contain oily substances which catch fire very readily. In **plantations** many trees of the same species are grown together and harvested, just like crops on a farm. The only way of stopping fires in these areas is to leave protective strips unforested as firebreaks.

To be really effective firebreaks have to be very wide because flames jump a long way when the fire is fanned by a strong wind. The difficulty is that a firebreak is not productive. Leaving the ground barren does not make any money. Most forests are not very well prepared in this way; the foresters take a gamble that there won't be a fire or that they will be able to deal with it.

Commercial forests often have watch towers placed so that foresters can spot a fire quickly. Teams have an emergency plan to ensure that many people can get to the scene of the fire as rapidly as possible.

► *This poster reminds people of the danger of fire and also of all the animals that can be harmed by an accidental fire in the countryside.*

40

◄ *"Smokey" was a small bear cub that was nearly burned to death in a forest fire. The National Forest Service now uses Smokey as a symbol of the dangers of fire.*

WHAT WILL IT TAKE?

Keeping people informed

The risk of fire is greatly increased by allowing visitors to enter a forest. Visitors may carelessly drop a lighted match or cigarette and start a fire. Many forests near public roads have a broad plowed strip next to the road to stop a fire being started by a lighted cigarette thrown from a passing car.

The best way to reduce the risk is to make people better informed of the danger and explain what will happen if a fire starts. Forest rangers often display a bulletin board explaining the risk of fire. When there has been a long period of drought the risk increases and if people are warned of the danger they will be extra careful in the woods.

Care in the City

Cities contain a multitude of places where fires might start. Fire disasters may begin in the home, at work, or in places where people are enjoying themselves such as a club, a restaurant, on the train, or in the car.

City fires often contain a greater variety of problems than those in the country. Warehouses store all types of materials that may present hazards. A fire may start in a building containing poisonous chemicals. There may be substances that can explode or give off poisonous fumes. Even in a hospital—usually a safe place—there are many gas tanks containing oxygen. If they become overheated they will explode with such force they could demolish the building.

Being prepared in the city is certainly a difficult task. It begins, however, with architects following the simple fireproofing rules.

How architects design for safety

Designing for safety (prevention) is much better than simply relying on efficient rescue. It is most important that buildings should be designed with safety in mind because people spend much of their lives inside buildings.

There are two ways of preparing against fire. The first is to use materials that are not combustible. The second is to prevent the fire from taking hold.

There is an enormous variety of materials used in any building. A great deal of research has been undertaken to find materials that are both well suited to their jobs and fire resistant.

Concrete or brick used in a building are usually safe in all but the worst fires. However, many houses are made entirely of wood. This is true in places as widely scattered as the United States and Canada, Austria and Norway, and throughout the developing world. Such houses pose a great problem, because they will start to burn in an intense fire. A wood building is also more likely to collapse than a brick building.

Building by compartments

Air is vital to life as well as to a fire, so you can't exclude air from places where people live and work. However, an architect can design his building to reduce the amount of air that gets to a fire. This is done by using **compartments,** spaces with walls.

A fire needs very large amounts of fresh air if it is to take hold strongly. The amount of air it needs is much greater than the air required by people for breathing. In buildings, the supply of air to a fire can be reduced by building using a design of closed compartments. An ordinary room makes a very good natural compartment. If every room in a building has its doors kept closed, a fire will not be able to get a large supply of oxygen and it will burn slowly. This will give time for the firefighters to arrive. Open area offices, department stores, hotels, and warehouses present special fire problems because there is a lot of air in each "compartment."

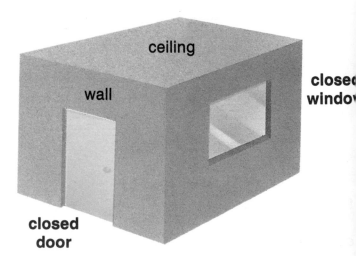

▲ *A compartment with closed doors and windows helps control a fire because no fresh air can get in.*

In these cases there are often metal fire barriers that will close automatically or can be closed by employees. Again their role is to turn the space into smaller compartments and so contain the fire.

Almost any form of door will stop a fire for several minutes, certainly long enough for people to get out of most buildings. There are special doors, called **fire doors,** which will resist a fire for over half an hour. These are used in critical places where it is important to keep fire at bay for as long as possible. Fire doors are placed at all entrances to a flight of stairs for example. This is because the stairs may be the only means of escape from a building.

An architect can also design for the safe evacuation of a building. There must be obvious ways to get out, called **escape routes.** It is very dangerous to have a building with many dead-end passages from which there is no escape.

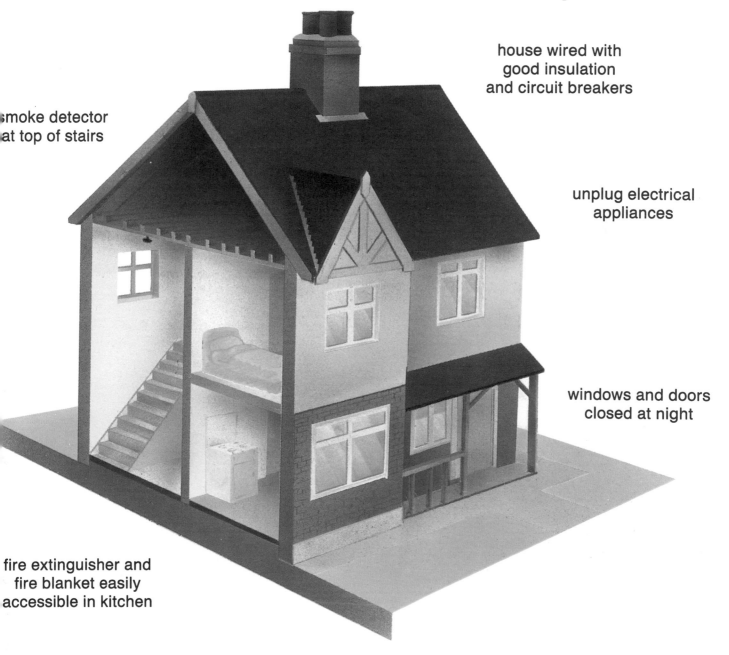

house wired with good insulation and circuit breakers

unplug electrical appliances

windows and doors closed at night

smoke detector at top of stairs

fire extinguisher and fire blanket easily accessible in kitchen

▲ *How to reduce the fire hazard in your own home.*

Helping people to escape

Public places present special risks because there are many people in unfamiliar surroundings. Having clearly marked exit routes is the first way to be prepared.

Tall buildings may be extra dangerous. If a fire starts in a lower floor people on the floors above may be trapped. To reduce this possibility carefully designed buildings have two well-separated staircases and an elevator. The elevator shaft and both staircases must be fireproof, with fire doors on all floors. Without this the fire will rush up the stair wells, sucking air in from below and using the stairs as a chimney. Instantly the fire will spread to many floors and people will have no escape.

> **Fire safety rules**
>
> 1. Set off any fire alarm so that others can escape quickly
> 2. Get out of danger quickly and without panic and then
> 3. Telephone for the firefighters and give them accurate information about the fire

Materials at home and at work

Although a building may be completely safe in design, the care that has gone into that design may be made useless by the way the building is used. A home is full of furniture and much of the upholstery may be of foam which is inflammable and gives off poisonous fumes. In an office there may be an enormous amount of paper which could readily catch fire. In a factory there may be machines that can overheat near chemicals that can then catch fire.

An architect cannot control what is kept in a building. An architect designs a building to try and contain any fire that might start. Safety also depends on people being careful in what they do, and in the materials they keep or use close to sources of heat. People need to know about these dangers—they must be taught fire awareness.

▼ *A hand-held fire extinguisher is useful to put out small fires. It must be easy to reach.*

◄ *A fire escape is a quick means of getting out of tall buildings. It provides an alternative escape route from the main staircase.*

▲ *A fire detector placed on a ceiling will set off an alarm and warn people at the first signs of smoke. These are now required in most homes and apartments.*

How people can be prepared

Everyone should know what to do and be prepared in case of fire. First and foremost it is important not to let any source of heat get to any combustible materials by accident. However, once a fire has caught, it is important not to create a chimney effect. If doors and windows are shut upon leaving a building, the chance of fresh air getting to a fire is reduced and safety is increased. If people do not understand the principle of compartments and leave doors open, or even prop open fire exit doors, lives will be put in danger.

Tackling a fire is a dangerous job. Although a fire extinguisher will put out very small fires, it will not have any effect on large fires. It is also important to use the right means to extinguish a fire. Throwing water on an oil fire will just make a bad situation worse. In any case, even a small fire can produce smoke and fumes that can kill.

How firefighters are prepared

All countries, and even some businesses, employ professional firefighters. Firefighters receive special training so they can tackle a job most effectively. To be prepared there must be people on duty all the time, all night as well as all day.

The firefighters must have good transportation. It is important to get to the scene of a fire as quickly as possible. Most fire trucks have certain basic equipment: a supply of water in a tank; hoses to connect to the fire hydrants in the street; ladders to rescue people who might be trapped; and other special extinguishers to deal with certain types of fires.

The firefighters must also protect themselves. They have to wear flame resistant clothes and have special breathing equipment so that they will not be overcome by the dense fumes that always accompany a fire.

◄　*Hoses, ladders, and fire engines.*

Glossary

acid rain
rainfall that contains dissolved gases from burning fuels. Acid rain kills trees and fish in lakes.

brush
small trees and bushes

bucket brigade
a group of people who work together to fight a fire by passing buckets of water to each other along a line

bush
wild forest and scrubland

carbon dioxide
a gas that absorbs heat from the Earth. The extra carbon dioxide carried into the atmosphere by fires is believed to have caused the air temperature to rise all over the Earth.

chimney
a vertical tube that carries hot air and smoke away from a fire

City Watch
the people employed in seventeenth century England as policemen

combustible materials
material that will catch fire when heated. Combustible materials include gasoline, wood, coal, paper, clothing, and some plastics.

compartment
a closed box. The idea of compartments is used by architects in designing buildings that help to keep fire contained.

desertification
the use of soil to such an extent that it is left unprotected and can be washed or blown away. In the end what remains is a sandy and stony desert.

developed countries
those countries that have a wide range of industries, and that make most of the world's wealth

developing countries
countries that have not yet fully industrialized and that do not have a wide range of health, water, and other facilities available to the majority of the people. In most developing countries the majority of the people work as farmers.

disaster
a severe event that changes the landscape or disrupts the normal lives of people

drought
a period without rain that is long enough to cause hardship to people, animals, and plants

dryland
a land that experiences a long dry season

embers
smoldering pieces of wood, coal, and other combustible materials

escape route
the route that has been designed to allow quick and safe escape. A fire escape route in a building is usually marked with signs saying "Fire Exit."

eucalyptus
a tall tree that grows quickly in regions with a long dry season. A eucalyptus tree contains many oily substances that cause it to burst into flame very easily.

fire apparatus
a term used by firefighters to describe the fire engine with all its specialist equipment for fighting fires

firebreak
a wide space containing no combustible materials and designed to stop the spread of a fire

fire department
a professional team of firefighters

fire door
a door made in such a way that it will not burn quickly. Its purpose is to allow plenty of time for people to escape from a burning building.

fire storm
an intense fire that causes violent winds to be sucked toward its center

flashover
the way trapped gases burst into flame if the gases are suddenly allowed to escape from a closed room into a place with air

flash point
the temperature at which a vapor catches fire

fossil fuel
the fuels such as coal, oil, and gas that were formed many millions of years ago and are now trapped underground

fuel
material that will burn and that can be used to give out heat energy

fuse
material that readily catches fire and is used to set other less combustible matter on fire

fuse wire
a thin piece of wire placed in an electrical circuit and designed to burn out safely if the circuit becomes overloaded. Each fuse is designed for a special purpose and must be used correctly.

hydrant
a standing pipe where firefighters can attach their hoses and draw water. Hydrants in some countries, such as the United States are in short posts on the sides of streets. In countries such as the United Kingdom they are in compartments set into the pavement.

incendiary bomb
a bomb containing materials that will spread fire when it explodes. An incendiary bomb often contains a substance called phosphorus which catches fire on contact with air.

incomplete combustion
when partial burning has taken place and a substance has become very hot. If more air is added to gases formed by partial combustion they will turn into a ball of flame.

insulation
the material around electrical wires that stops electricity flowing between the wires. In a cable the insulation is usually some form of plastic. In older wiring it was rubber and cotton.

nutrients
the chemical foodstuffs that plants absorb from rain and soil water

plantation
a large commercial "farm" that grows only a few types of trees or other crops

pollute
make dirty or contaminate with materials that cause a danger to health. Pollutants in the air include soot and gases released by fire.

prairie
grasslands situated in the central regions of the United States and Canada

rodent
a small furry animal with large front teeth

savanna
a region of tropical grasslands and short trees found where there is a long dry season

spontaneous
happening without any outside influence. Spontaneous combustion occurs without a source of flame.

steppe
another word for prairie used to describe grasslands of the Soviet Union.

suffocation
death caused by lack of air

tinder
dry wood or other combustible material used for lighting a fire

vapor
gas given off by a liquid

Index